YOUR KNOWLEDGE HAS VALUE

- We will publish your bachelor's and master's thesis, essays and papers

- Your own eBook and book - sold worldwide in all relevant shops

- Earn money with each sale

Upload your text at www.GRIN.com and publish for free

Ataliba Miguel

Well Planning at Molly Field

GRIN Verlag

Bibliografische Information der Deutschen Nationalbibliothek:

Die Deutsche Bibliothek verzeichnet diese Publikation in der Deutschen National-
bibliografie; detaillierte bibliografische Daten sind im Internet über http://dnb.d-
nb.de/ abrufbar.

Imprint:

Copyright © 2012 GRIN Verlag GmbH
Druck und Bindung: Books on Demand GmbH, Norderstedt Germany
ISBN: 978-3-656-71301-2

This book at GRIN:

http://www.grin.com/en/e-book/278037/well-planning-at-molly-field

Well Planning Molly Field

Prepared By:

Ataliba Miguel

June 2012

Word Count 3034

Summary

A new prospect is planned to be drilled by Kenmac Petroleum Corporation on a new licensed Molly field located in the North Sea on Block 14a/b. The tasks involved in well planning are as follows: (a) the estimation and evaluation of pore and fracture pressure and their impact on well planning, (b) a lithology study of the different formations and the steps taken to minimize the occurrence of drilling problems through all different formations, (c) a casing scheme design, (d) a mud programme design for the given well, and (e) a cement type suitable for the application in the given well.

List of Contents

List Figures

- 4 -

List of Tables

1. Introduction

A new prospect is planned to be drilled on a new exploratory well at the Molly field. The field is located in the North Sea Block 14a/b. Kenmac Petroleum Corporation has been assigned to carry out a number of tasks necessary as a part of an effective well planning. Feasibility studies based on pore and fracture will be evaluated in order to assess the possible zones of abnormal formations and the steps necessary to prevent drilling problems. Furthermore, a casing scheme will be proposed along with the mud programme and the cementation suitable for the particular well.

1.1. Objectives

- Drill vertically a new exploratory well
- Evaluate reliable data source for pore and fracture pressure
- To elaborate a casing scheme, a mud programme and the
Cementation

2. Pore Pressure

Pore pressure is the pressure exerted by the fluids contained in the pore space of a rock. It is normally related with the fluid column density and vertical depth. Pore pressure data can be predicted utilizing (1) seismic interval velocities, and (2) offset well logs (Selim, Badway and Abdullah, 2010). Wildcat well planning, often uses the predictive method to estimate the formation pressure. Predictive methods are primarily based on (1) data available from nearby wells and (2) seismic data.

2.1. Estimating Pore Pressure

Two basic approaches are used to make a quantitative estimate of formation pressure (Bourgoyne et al, 1991, p. 253)

- Plots of a porosity-dependent parameter vs. depth – assuming that the effective stress matrix throughout an abnormally pressured formation as well in normally shallow pressured formation, are the same.
- Plots of a porosity-dependent parameter vs. depth using empirical correlations.

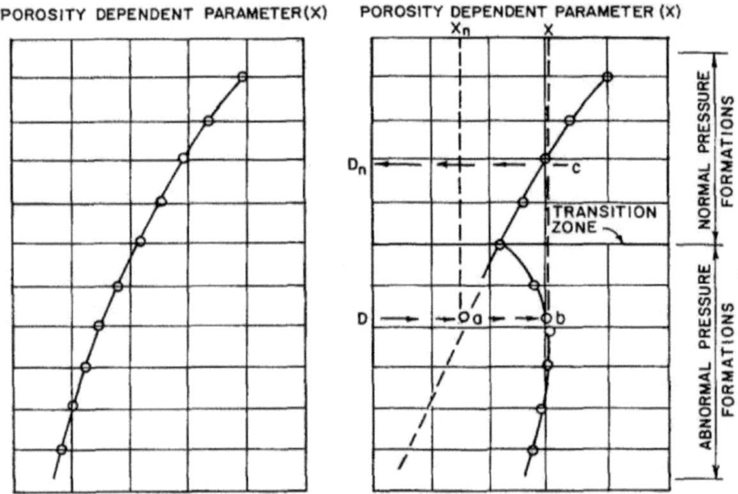

a. Normally Pressured Formations   b. Abnormally Pressured Formations

Fig. 1 – Normal and Abnormal pressure graphs (Adapted from Bourgoyne et al, p.254)

Formation pore pressure can be normally or abnormally pressured. The normal pore pressure is the hydrostatic pressure due to the average density and vertical depth of the column of fluids above a particular point (Bera, 2010, p.1). As such pressures higher than normal are termed geopressured or over pressured and pressures lower than normal are termed subnormal, table below gives some typical values.

Normal	Abnormal	
Formation Pressure (psi/ft)	Formation Pressure (psi/ft)	
	Subnormal	Geopressured
0.433	< 0.433	> 0.465

Table 1 – Typical values for normal and abnormal pressure formations

Bourgoyne et al (1991) have clearly summarized four mechanisms for abnormal pore pressures, and these are:

- ➢ Compaction
- ➢ Diagenesis
- ➢ Differential density and
- ➢ Fluid migration

The most common abnormal pore pressure generating mechanism is the compaction.

2.2. Formation Pore Pressure - Importance

The prediction of pore pressure is of paramount importance in the well planning, because it helps to maximize the drilling safety, the borehole stability, the rig selection (Bera, 2010), anticipate the location of potential abnormal pressure formation zones, as well as helps to minimize the drilling cost, so that the mud density can be optimized to provide sufficient overbalance and assure the casing selection with depth, will withstand the various formation pressures (Selim, Badway and Abdullah, 2010). The wellbore pressure should always as reasonable as possible be kept at a pressure higher than the formation pressure. The wellbore pressure scenario mostly desirable when planning a well is the overbalanced pressure. The aim is to prevent the influx of formation fluids into the wellbore and avoid possible kicks.

2.3. Sources of Pore Pressure

When formation pressures are normal, the porosity dependent parameter should have an easily recognized trend, because of the decreased porosity with increased depth of burial and compaction. A departure from the normal pressure trend signals into a probable transition to abnormal pressure is very crucial for the well planning and drilling operations, since the depth at which the departure occurs is critical because the casing must be set in the well before excessively pressured permeable zones can be drilled safely (Bourgoyne et al, 1991, p.253).

Table 2- Molly Field – Geology Prognosis

Series	Depth Interval (ft)	Depths (ft)	Lithological Description
Paleocene	Sea bed to 5910	Sea bed - 105	Silts & soft clays
		105 - 650	Unconsolidated sands
		650 - 4260	Paleocene middle shales, clays and light sandstones
		4260 - 5400	Paleocene lower sands
		5400 - 5910	Danian, chalks and fine/medium
Cretaceous	5919 - 7260	5919 - 7260	Upper and lower Cretaceous chalk/limestones
Jurassic	7260 - 7980	7260 - 7980	Upper Jurassics, shales/sandstones leading at lower end to fine/medium "Piper" sandstones
Triassic	7980 - 9765	7980 - 9765	Triassic, shales and possible fluid bearing sandstones on post Zechsteins
Permian	9765 - 9960	9765 - 9960	Permian Zechsteins, anhydrite, and dolomites
Carboniferous	9960 - 10000	9960 - 10000	Carboniferous siltstones/shales

The prognosis depths and pore pressures for the reservoir formations are listed in the table below.

Table 3 – Pressure Gradient versus Depth

Formation	Depth Interval (ft)	Pressure Gradient	
		psi/ft	ppg
Seabed to Paleocene	4260	0.465	8.9525534
Paleocene to Jurassics	3720	0.558	10.7430641
Jurassic to TD	2020	0.614	11.821221

2.4. Fracture Pressure

Fracture pressure is the critical pressure required to break down the formation or induce fractures. For well planning fracture gradient is used as a plot of pressure vs. depth to estimate when a rock will fail and induce fractures.

Formation fracture data are estimated by predictive methods. Since formation fracture pressure is affected greatly by the formation pore pressure (Bourgoyne et al, 1991), the methods of correlation applied in pore pressure prediction, must be applied before using a fracture pressure correlation. Fracture gradient helps to determine the setting depths for intermediate casing strings, the maximum allowable annular surface pressure allowed to control a kick, and maximum allowable mud density for drilling.

2.5. Methods for Estimating Fracture Gradient

Fracture pressure is estimated by using equations and correlations.
Equations are from Hubbert and Willis, and Christman. The correlations are from Mathew and Kelly, Penabaker, Macpherson and Berry, and Eaton.

3. Drilling

Several problems are encountered while drilling through different formations. A study of lithology column helps to predict the possible zones of drilling problems and as so the measures to deal with those problems. The following paragraphs explain the background of drilling through different formations.

During drilling operations, the bottomhole pressure should be maintained slightly higher than the formation pore pressure. The confinement associated with this static differential pressure causes an increase in the drilling strength of rock. During drilling, the rock ahead of the cutter is rapidly deformed. The rock deformation associated with the dilatation leads to a dynamic pressure differential between the borehole fluid and pore fluids of the rock being cut that can equal the total borehole pressure. Dynamic confinement pressure increases the strength and plasticity of rock, reducing the efficiency of indentation and shear cutting
(Kolle., 2000. p. 1).

3.1. Rate of Penetration

The rate of penetration is considered as one of primary factors which affect the drilling efficiency through different lithology formations. As such bit torque is used to classify lithology into three categories, (1) porous, (2) argillaceous (shaly), and (3) tight (Wolcott and Bordelon, 1993, p.769).
Porous rock e.g. sandstones are weaker than argillaceous rocks (shales), which in turn are weaker than tight rocks (Falconer, Burgess and Sheppard, 1998 p.124).
The problems that may be encountered while drilling through different formations are related to a transition between normal pressure zones to over pressured zones. The rate of penetration is strongly affected by some important variables such as (1) bit type, (2) formation characteristics, (3) drilling fluid properties, (4) bit operating conditions (bit weight and rotary speed), (5) bit tooth wear and (6) bit hydraulics.

3.2. Rock Mineralogy

The mineralogy of the rock has some effect on penetration rate. Rocks containing hard, abrasive minerals can cause rapid dulling of the bit teeth. Rocks containing gummy clay minerals can cause the bit to ball up and drill in a very inefficient manner.

3.3. Lithology Study

The Molly Field lithology column is characterized by sandstones, shales and clay.

3.4. Sandstone zone

Drilling through hard sandstone poses particular problems for bit optimization. Conventional methods for raising rate of penetration ROP, such as use of PDC bits, are limited by rapid cutter wear in hard, quartz rich formations. In these non hydratable formations, bit cleaning may also exert a negligible effect on rate of penetration ROP, closing off another potential source of improvement. In practice this means that options to raise rate of penetration ROP in hard sandstone may be restricted to exploiting the effects of drilling parameters, with bit types capable of resisting rapid abrasive wear (Fear, 1999, p. 46)

3.5. Shale zone

When drilling through shales formations the two major problems encountered are the shale sloughing and swelling. Shale instability in a wellbore is attributed to any of the following combination of forces: (i) overburden pressure, (ii) degree of compaction at the formation, (iii) pore pressure in shale exceeding the hydrostatic pressure (iv) tectonic forces and (v) presence of micro fractures along cleavage planes on the clay platelets (Talabani, Chukwu and Hatzignatiou, 1993, p.283).

Brittle (sloughing) shale, Gumbo (plastic) shale and Hydratable (swelling) shale are the major types of shale usually encountered when drilling a borehole. Shales are high compressible sedimentary rocks possessing the ability to absorb water through their lattice structure. They are composed of different minerals such as quartz, feldspar, dolomite, calcite, siderite and gypsum. These minerals are considered inert and are not affected by the drilling fluids. Other minerals such as kaolinite, illite, chlorite,

montmorillonite and mixed layer clays absorb water when exposed to water base fluids creating various degree of instability (Talabani, Chukwu and Hatzignatiou, 1993, p.284).

3.6. Drilling problems

Typical problems with shale sloughing are: pipe stuck when drilling through a depth interval, heaving of the shale, loss of circulation, very fast drill rotation string. This in turn creates borehole enlargements.

One of several methods used to deal with these problems is to change the mud system in place and increase the mud equivalent circulating density (Talabani, Chukwu and Hatzignatiou, 1993, p.285).

4. Casing Programme

The casing string scheme and their respective setting depths are based on pore and fracture gradient of the formations to be penetrated
(Bourgoyne et al. 1991. p.330).

The size of the casing string is controlled by the necessary internal diameter ID of the production string and the number of intermediate casing strings to reach the target depth (Bourgoyne et al. 1991. p.331). The design scheme is typically made from bottom to top.

The Molly field casing scheme will be adopted as follow:
> Production tubing
> Production casing string
> Intermediate casing string
> Surface casing string
> Conductor casing string

4.1. Conductor Casing

The conductor casing act as a guide for the remaining casing strings into the hole. It serves also to cover unconsolidated formations and to seal off over pressured formations. The outside diameter size to be used in the casing scheme is 30 in.

4.2. Surface Casing

The surface casing is set deeply enough to protect the borehole from caving-in in loose formations frequently encountered at shallow depths, and protects the freshwater sands from contamination while subsequently drilling a deeper hole (Lyons and Plisga, 2005, p.406). The outside diameter size to be used in the casing scheme is 16 in.

4.3. Intermediate Casing

The intermediate casing is usually set in the transition zone before abnormally high formation pressure is encountered, to protect weak formations or to case off loss-of-circulation zones (Lyons and Plisga, 2005, p4-406). The outside diameter to be used in the casing scheme is 10 ¾ in.

4.4. Production Casing

The production casing is the last string to be run in the wellbore and is either run through the pay zone (reservoir) or set just above the pay zone for an open hole completion. The main purpose of production casing is to isolate the production interval from other formations such as water bearing sands, and to protect the completion tubing. The outside diameter to be used 7 5/8 in.

4.5. Production Tubing

Tubing string is installed in the hole to protect the casing from erosion and corrosion and to provide control for production. (Lyons and Plisga, 2005, p4-467). The outside diameter to be used in the casing scheme is 5 in.

Table 4 – Casing Sizes

	Casing					Depth (ft)
	Grade	Yield Strength (psi)	Tensile Strength (psi)	OD (in)	Hole (in)	BML (below mud line)
Conductor	K-55	55,000	75,000	30	36	980
Surface	K-55	55,000	75,000	16	20	4250
Intermediate	K-55	55,000	75,000	10 3/4	14 3/4	8000
Production	K-55	55,000	75,000	7 5/8	9 1/2	10000

4.6. Casing Setting Depth

The fracture gradient is an essential element in the selection of the casing setting depth. At shallow depths, measurements from LOT (leakoff test) reveal that fracture pressure gradients are large at shallow depth, but steadily decreases with depth. Shallow casing strings provide little protection against shallow gas kicks (Aadnoy et al, 1989, p. 827).

Since it has been assumed that no surface or shallow gas are expected in the Molly field area, then shallow gas kick pose no problems on the conductor casing shoe.

From data given the pressure gradient at 10,000 ft is 0.614 psi/ft (equivalent mud weight = 11.8 ppg). In order to control the this pressure, the wellbore pressure must be greater than 0.614 psi/ft. Typically a factor between 0.025 to 0.045 psi/ft is used to take into account the effects of swab and surge and provide a safety factor (Rahman and Chilingarian, 1995, p.123). Thus the pressure gradient required to control formation pressure at the bottom of the hole would be 0.614 + 0.025 = 0.639 psi/ft (12.28 ppg equivalent mud weight).

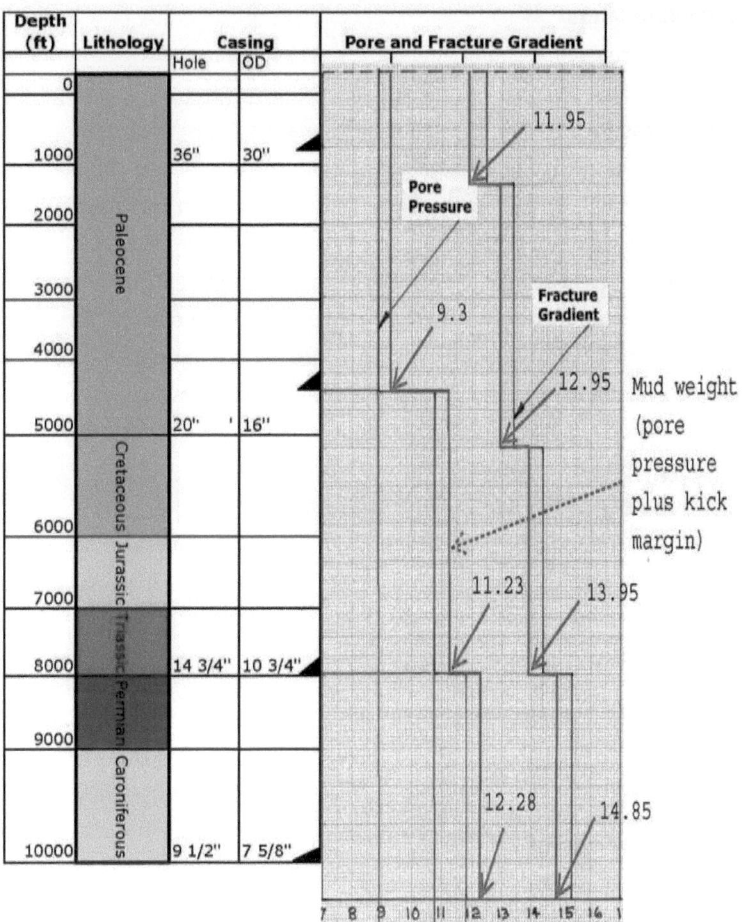

Fig. 2 – Casing setting depth

5. Drilling Muds

Drilling fluids are used in the drilling process (1) to clean the rock fragments from beneath the bit and carry them to the surface, (2) to control the formation pressure avoiding the influx of formation fluids from flowing into the well, (3) to prevent the hole from caving in, (4) to cool and lubricate the bit and drill string, (5) to allow easy formation evaluation and to minimize corrosion of the drill string.

The drilling fluid selections are mainly governed by (1) the types of formations to be drilled, (2) the range of temperature, strength, permeability, and pore fluid pressure exhibited by the formations, (3) the formation evaluation procedure, (4) the water quality available, and (5) ecological and environmental considerations.

5.1. Basic Mud Properties and Ingredients

Five basic properties are usually defined by the well program and the monitored while drilling (Schlumberger, 1994). These are as follows:

> Rheology
> Density
> Fluid loss
> Solids content
> Chemical properties

5.2. Water Base Mud

Consist of a mixture of solids, liquids, and chemicals, with water being the continuous phase (Bourgoyne et al, p.41). Water base drilling fluids normally contains a number of different types of clays. Clays are added to attain some desirable physical properties and eliminate hole problems.

The commonest types of clays incorporated into the drilling fluid are calcium montmorillonite (also known as Wyoming Bentonite), illites and kaolinites. Bentonite is the most used as a drilling fluid. It is added to the water base drilling fluids to increase the viscosity and gel strength of the fluid. The most important function of bentonite is to improve the filtration cake properties of the water base drilling fluid (Lyons and Plisga, 2005, p. 110).

5.3. Water Base Mud Types

- ➢ Spud Muds – generally prepared with water and bentonite. Though it is untreated chemically, lime, cement and caustic soda is added to increase viscosity and give the mud a fluff to seal possible lost return zones. Spud muds are used for drilling the surface hole (Lyons and Plisga, 2005, p. 112)

- ➢ Natural Mud – generally use native drilled solids incorporated into the mud for viscosity, weight, and fluid loss control. They are often supplemented with bentonite for added stability and water loss control. Natural muds are generally used in top hole drilling to mud-up or to conversion depth.

- ➢ Saltwater Muds – are classified as salt water muds when they contain more than 10,000 mg/l of chloride. They may also be further classified according to the amount of salt present and/or source of makeup water. Saltwater muds include the following types: (1) sea water or brackish water muds, (2) saturated salt muds, (3) chemically treated mud, (4) lignite mud, (5) calcium treated muds, (6) lime muds, and (7) gypsum muds.

5.4. Oil Base Mud

Oil base muds contain oil as the continuous phase and trace amounts of water as the dispersed phase, emulsifiers, wetting agents and gellants.

The oil used for an oil base mud can be diesel oil, kerosene, fuel oil, selected crude oil, mineral oil, vegetable esters, linear paraffin, olefins, or blends of various oils. Emulsifiers are very useful in oil base mud due to the water contamination on the drilling rig. The stability of an emulsion mud is an important factor that has to be closely monitored while drilling.

Poor stability results in coalescence of the dispersed phase and the emulsion will separate into two distinct layers.

Oil base muds are well suited for drilling in (1) troublesome shales that swell and disperse, (2) in deep high temperature holes in which water base muds solidify, (3) in water soluble formations such as salt, anhydride, camallite, and potash zones, and (4) in the producing zones.

Additionally oil base muds can be used as:
- Completion and workover fluid
- A spotting fluid to relieve stuck pipe
- A packer fluid or a casing pack fluid

Table 5 - Mud programme

Depth (ft)	Formation Fluid Gradients (psi/ft)	Equiv. Mud Weight (ppg)	Mud Weight + trip margin (ppg)
0 - 980	0.465	8.9	9.3
0 - 4260	0.465	8.9	9.3
0 - 7980	0.558	10.72	11.22
0 - 10000	0.614	11.78	12.28

6. Cement

Cement is used in the drilling operations to protect and support the casing, to prevent the movement of fluid through the annular space outside the casing and to stop the movement of fluid into vugular or fractured formations (Bourgoyne et al, 1991, p. 85).

The cement slurry should have adequate compressive strength and low permeability when the cement hardens. The critical factor in obtaining a satisfactory cement job is to place the cement completely around the casing to prevent channeling.

6.1. Cement Classification

Portland cement is main ingredient in almost all drilling cements. It is artificial cement made by burning a blend of limestone and clay.

API has defined eight standard classes and three standard types of cement for use in wells. The eight classes are designated Class A to Class H

(Bourgoyne et al, 1991, p.89). Oil well cements are basically classified by the distribution of five basic compounds which are used to make cement: (1) C3S, (2) C2S, (3) C3A, (4) C4AF and (5) CaSO4

The API approved classes of cement are:
- Class A and B: generally used at shallow depths. Class B has a higher resistance to sulphate than Class A.
- Class C: it is used also at shallow depths, when conditions require high early strength.
- Classes D, E and F: are used at relatively higher depths. Classes D, E and F are known as retarded cements due to a coarser grind, or the inclusion of organic retarders.
- Classes G and H: Are the general purpose cements which are compatible with most additives. Class G is the common type of cement used in most areas.

6.2.   Cement Selection

The suitable type of cement to be used in the different sections of the well in the Molly field is the Class A, G and H. The reason behind this selection lies in the cost involved versus depth, and the wide range of well depths and temperatures that they cover.

The production string is the longer casing string, and as such the cement job is carried out in two stages. The production string is only cemented at the bottom where the cement slurry does not fill the entire annulus, reaching only a certain height within the formation. The objective of this partial cementation is that at some point production string will have to be removed and replaced, so cementing back to surface will have some negative impact in the operations of replacing the production casing.

7. Conclusion

The new licensed Molly field is comprised with zones of normal and abnormal pressure. The pore pressure and fracture pressure have played an important in the determination of casing setting depths and casing shoes. Furthermore, some drilling problems where encountered due to different lithology formations on sandstones and shales. The mud program was designed taking into account the values of pore and fracture pressure, kick margin tolerance to prevent the influx of formation fluids.

The cementation was adopted taking into account the cost and its wide range of well depths and temperature they cover.

References

Aadnoy, B.S. et al., (1989) 'Casing Point Selection at Shallow Depth'.
SPE/IADC Drilling Conference. February 28 – 03 March. Society of
Petroleum Engineers. pp. 825-836

Aadnoy, B. S., Rogaland, U. and Larsen, K. (1989) 'Method for Fracture-
Gradient Prediction for Vertical and Inclined Boreholes'. *Journal of
Drilling Enginering,* 4(2), pp. 99-103. [Online] Available at
http://www.onepetro.org/mslib/app/Preview.do?paperNumber=0001669
5&societyCode=SPE
(Accessed on 01 June 2012).

Bera, P. (2010) 'Estimation of Pore Pressure from Well Logs: A theoretical
analysis and case study from an offshore basin, North Sea'. *8th Bienial
International Conference and Exposition on Petroleum Geophysics.*
[Online]. Available at: http://www.spgindia.org/2010/217.pdf (Accessed
on 15 June 2012)

Falconer, I.G., Burgess, T.M. and Sheppard, M.C. (1998) 'Separating Bit
and Lithology Effects from Drilling Mechanics Data'. *IADC/SPE Drilling
Conference.* Dallas, Texas, February 28 – 02 March, Dallas, Texas.
Society of Petroleum Engineers. pp. 123-136

Fear, M.J. (1999) 'How to Improve Rate of Penetration in Field
Operations'. *IADC/SPE Drilling Conference.* New Orleans, 12-15 March.
New Orleans. Society of Petroleum Engineers. p.p 42-49

Kolle, J.J, (2000) 'Increasing Drilling Rate In Deep Boreholes by
Impulsive Depressurization'. [Online]. Available at:
http://mmstbpi.weebly.com/uploads/4/7/2/5/4725854/5331_16_rock_m
echanics.pdf [online], accessed on May 18

Selim, S. M., Badaway, B. E. and Abdullah, K. M. (2010) ' Pore Pressure
and Pressure Regime Evaluation, Role and Contribution in Well Planning
and Formation Evaluation Process, Gulf of Suez Oil Fields – Egypt'. *North*

Africa Technical Conference and Exhibition. Cairo, Egypt, 14-17 February. Cairo. Society of Petroleum Engineers. pp. 1-7

Talabani, S., Chukwu, G. and Hatzignatiou, D. (1993) 'Drilling Successfully Through Deforming Shale Formations: Case Histories'. *Rocky Mountain Regional/Low Permeability Reservoirs Symposium.* Denver, USA, 12-14 April. Society of Petroleum Engineers. pp. 283-289

Warren, T. M. and Smith, M. B. (1985). 'Bottomhole Stress Factors Affecting Drilling Rate at Depth'. *Journal of Petroleum Technology,* 37(8), pp. 1523-1533. [Online] Available at: http://www.onepetro.org/mslib/servlet/onepetropreview?id=00013381# (Accessed on 17 June 2012)

Wolcott, D. S. and Bordelon, D. R. (1993) 'Lithology Determination Using Downhole Bit Mechanics Data'. *68th Annual Technical Conference and Exhibition.* Houston, USA, 3-6 October. Houston. Society of Petroleum Engineers. pp. 769-778

Bibliography

Bourgoyne et al. (1991). Applied Drilling Engineering. Richardson, Texas. Society of Petroleum Engineers.

Lyons, W. C. and Plisga, G. J. (2005). Standard Handbook of Petroleum and Natural Gas Engineering, 2nd Ed. Massachusetts, USA. Elsevier.

Rahman, S.S. and Chilingarian, G.V. (1995). Casing Design Theory and Practice. Amsterdam,. Elsevier.